Copyright Page

This book is self published in electronic form, and the entire content is copyrighted

By Martin K. Ettington All Rights Reserved USA 2008
Printed in the United States of America

On Using the Scientific Method for the Paranormal

The Crazy and Out of the Box Series Book 3

First Edition 2000
Second Edition 2008
Third Edition 2011

ISBN: 9781976921018

Library of Congress Cataloging-in-Publication Data

1) Superscript references of [B2] or [W5] for examples refer to the Bibliography or Web Links references at the end of the book

2) If you the reader are interested in contacting the author he can be reached at the email address: marty@personal-longevity.com

3) This book is self published in electronic form, and the entire content is copyrighted by Martin K. Ettington All Rights Reserved USA 2000

4) This second edition was done eight years after the original was written. A lot of personal info and non relevant material was removed and the whole book was reformatted using Word 2007.

On Using the Scientific Method to Study the Paranormal

"On Using the Scientific Method to Study the Paranormal" is an analysis of how the un-measureable can be measured.

The basis of Science is the ability to test a Hypothesis.

This can only be done by having instruments which can measure the phenomena in question. If measurements can't be made then Science can't be conducted.

After discussing some of his personal history with Paranormal experiences, the Author proposes some ideas to allow scientific validation of the paranormal which is normally a very subjective experience.

Concepts of Reality and how Science evolved are discussed.

Other books by Martin K. Ettington

Spiritual and Metaphysics Books:
Prophecy: A History and How to Guide
God Like Powers and Abilities
Enlightenment for Newbies
Removing Illusions to Find True Happiness
Using the Scientific Method to Study the Paranormal
A Compendium of Metaphysics and How to Guides (Six books together in one volume)
Love from the Heart
The Enlightenment Experience
Learn Your Soul's Purpose
Pursuing Enlightenment
A Modern Man's Search for Truth
Use Intuition and Prophecy to Improve Your Life
The Handbook of Spiritual and Energy Healing

Longevity & Immortality:
Physical Immortality: A History and How to Guide
The Commentaries of Living Immortals
Records of Extremely Long Lived Persons
Enlightenment and Immortality
Longevity Improvements from Science
The 10 Principles of Personal Longevity
Telomeres & Longevity
The Diets and Lifestyles of the Worlds Oldest Peoples
The Longevity Six Books Bundle

Science Fiction:
Out of This Universe
Personal Freedom-Parts 1 & 2
The Psychic Soldier Series:
  Book 1-Himalayan Journey
  Book 2-A Soldier is Born
  Book 3-Fighting For Right
  Book 4-Earth Protector
The Immortality Sci Fi Bundle

The God Like Powers Series:
Human Invisibility
Invulnerability and Shielding
Teleportation
Psychokinesis
Our Energy Body, Auras, and Thoughtforms

The God Like Powers Series—
  Volume 1 Compilation
The Yoga Discovery Series:
Yoga-An Ancient Art Form
Hatha Yoga-Helping you Live Better
Raja Yoga-Through the Ages
The Yoga Discovery Package

Business & Coaching Books:
Creating, Publishing, & Marketing Practitioner Ebooks
Building a Successful Longevity Coaching Business
Why Become a Coach?
The Professional Coaching Success Trilogy
2020-Make Money Writing and Selling Books
The 2020 Handbook of High Paying Work Without a College Degree

Science, Technology, and Misc.
Future Predictions By and Engineer & Seer
The Unusual Science & Technology Bundle
The Real Atlantis-In the Eye of the Sahara
Are Cryptozoological Animals Real or Imaginary?
Real Time Travel Stories From a Psychic Engineer
Removing Limits On Our Consciousness-And Thinking Outside the Box
33 Incredible True Survival Stories
How to Survive Anything: From the Wilderness to Man Made Disasters
All About Mars Journeys and Settlement
Mining the Asteroid Belt

Ancient History
The Real Atlantis-In the Eye of the Sahara
Ancient & Prehistoric Civilizations
Ancient & Prehistoric Civilizations-Book Two
The History of Antediluvian Giants
The Antediluvian History of Earth
Ancient Underground Cities and Tunnels
Strange Objects Which Should Not Exist
Strange and Ancient Places in the USA
A Theory of Ancient Prehistory And Giant Aliens

Aliens and Space
Aliens and Secret Technology
Aliens Are Already Among Us
Designing and Building Space Colonies
Humanity and the Universe
All About Moon Bases
All About Mars Journeys and Settlement
The Space and Aliens Six Books Bundle
A Theory of Ancient Prehistory and Giant Aliens
The Space Colonies and Space Structures Coloring Book
All About Asteroids

The Longevity Training Series

(A transcription of the online Multimedia Longevity Coaching Training Program)

The Personal Longevity Training Series-Book1-Long Lived Persons
The Personal Longevity Training Series-Book2-Your Soul's Purpose
The Personal Longevity Training Series-Book3-Enable Your Life Urge
The Personal Longevity Training Series-Book4-Your Spiritual Connection
The Personal Longevity Training Series-Book5-Having Love in Your Heart
The Personal Longevity Training Series-Book6-Energy Body Health
The Personal Longevity Training Series-Book7-The Science of Longevity
The Personal Longevity Training Series-Book8-Physical Body Health
The Personal Longevity Training Series-Book9-Avoiding Accidents
The Personal Longevity Training Series-Book10-Implementing These Principles

The Personal Longevity Training Series-Books One Thru Ten

These books are all available in digital and printed formats from my website and on Amazon, Barnes & Noble, Apple ITunes, and many other sites

My Books Website is: http://mkettingtonbooks.com

Signup for our Mailing List to get the following:

1) A discount coupon for 25% discount on all books on our site

2) Occasional Notices of new books available

3) Occasional Email on other offerings of ours (Monthly)

Go to this link to sign-up:

http://personal-longevity.com/mkebooks/emailsignup/

And click this link to get the FREE 102 page Ebook titled "Secrets of Many Things"

If you have any questions about this book or other subjects please contact the Author at:
mke@mkettingtonbooks.com

On Using the Scientific Method to Study the Paranormal

On Using the Scientific Method to Study the Paranormal

Acknowledgements

In the first version of this book Dr. Joaquin Fuster did very helpful reviews on the material.

This version of the book is Revision 1 based on feedback from different reviewers.

I would particularly like to thank my cousin Debbie Ettington and a distant relative Hal Atkins for their useful feedback which has allowed me to improve this work immensely

On Using the Scientific Method to Study the Paranormal

On Using the Scientific Method to Study the Paranormal

Dedication

This book is dedicated to my friend Dr. Sam Lentine who was my philosophical and meditation mentor on many concepts about the psychic/non physical realm.

Sam died in 1991, and although he was a pioneer researcher in the physics of mental phenomena and psychotronics which deals with the practical understanding of psychic energies, he never lived to see wide acceptance of what he believed.

I hope this book will contribute to the acceptance of ideas I know he would agree are important to the world.

On Using the Scientific Method to Study the Paranormal

On Using the Scientific Method to Study the Paranormal

Table Of Contents

Introduction	13
Chapter 1: An Overview of this Book	17
A. SOME MAJOR DILEMMAS OF HUMANITY	17
B. MY POINT OF VIEW	17
Chapter 2: A Summary of My Beliefs about Reality	21
	21
Chapter 3: Some Relevant Experiences from my Personal History	23
EXPERIENCES AS A TEENAGER	23
A. AGES 11-15 DID ESP EXPERIMENTS	23
B. FIRST CONSCIOUS OUT OF BODY EXPERIENCE	24
C. 1974-1975 EXPERIMENTS IN KIRLIAN PHOTOGRAPHY	25
D. MY PSYCHIC MENTOR	27
E. A VISION WHICH CAME TRUE	29
F. READ A LOT MORE METAPHYSICS	31
G. COURSE ON PSYCHIC PHENOMENA AT RPI	32
H. A FRIGHTENING EXPERIENCE WITH AN INVISIBLE ENTITY	34
I. SOME MEDITATION EXPERIENCES	36
EXPERIENCES AS AN ADULT	39
J. MENTAL WARNING OF A MUGGING	39
K. WARNING OF A DISASTER	41
Chapter 4: THE SCIENTIFIC METHOD	43
A. WHAT IS THE SCIENTIFIC METHOD?	43
B. THE NON SCIENTIFIC METHOD OF OBSERVATION	47
C. SCIENCE VERSUS PSEUDOSCIENCE	48
D. SUBJECTIVE VERSUS OBJECTIVE KNOWLEDGE	50
E. CAN EVERYTHING BE MEASURED OBJECTIVELY?	50
Chapter 5: MY PHILOSOPHICAL EVOLUTION	53
A. ARE PSYCHICS MENTALLY ILL?	53

On Using the Scientific Method to Study the Paranormal

B. WHY DO WE SEARCH FOR THE MEANING OF LIFE? 55
C. HOW MY THINKING HAS EVOLVED 58
Chapter 6: TOWARDS A BETTER FRAMEWORK OF REALITY 59
A. OUR LIMITED UNDERSTANDING OF REALITY 59
B. THE SCOPE OF REALITY 60
C. THE SCALE OF BELIEVABILITY 63
D. THE "REAL REALITY" 66
E. FANTASY AND REALITY ARE OBVIOUSLY DIFFERENT 68
F. ON THE POSSIBLE NATURE OF CONSCIOUSNESS AND REALITY 69
G. WHAT IS THE FORCE OF WILL? 71
H. THE PROBABILITY OF THE FUTURE 73
Chapter 7: WHAT THE FUTURE CAN HOLD 76
A. THE CONFRONTATION OF SCIENCE AND SPIRITUALITY 76
A. A SIMPLE NEUTRAL ANALYSIS: 77
B. A DEEPER NEUTRAL ANALYSIS: 77
C. A SYNTHESIS OF SCIENCE AND SPIRITUALITY? 79
D. A ROADMAP TO A NEW WAY OF THINKING 81
E. A MODIFIED SCIENTIFIC METHOD 83
F. DEVELOP A NEW OPENNESS TO REALITY: 85
G. WHY MANY SCIENCE MAY BE MISSING SOME OF THE TRUTH 87
Chapter 8: Recommendations for Action 90
Bibliography 94
Web Reference Links 96
Glossary of Words and Terms Used 98
Index 105

On Using the Scientific Method to Study the Paranormal

Introduction

A wise man once said that every person's life has enough interesting in it to fill at least one book.
This book is an outgrowth of my life experiences, and I feel I have a unique and valuable contribution to make to humanity's views of Reality and how Science and Religion can better work together.

In the book Proverbs 1:10 of the Bible it says that "Nothing under the sun is new, neither is any man able to say: Behold this is new: for it hath already gone before in the ages that were before us." [B3]. However, if the Second Millennium taught us anything, it's that man has only touched the beginning of unraveling the mysteries and complexities of our universe.

I really believe that today humanity has advanced to one of those thresholds where a new paradigm is needed to allow for radical new breakthroughs in the sciences and in people's understandings of themselves and how they fit into Reality.

This book is intended to help open people's minds to new concepts about Reality, and new approaches to understanding Reality.

This book is organized into five main sections:

Chapters 1-2- In the first Section of the book I cover some of my concerns from my point of view about how Science and Spirituality co-exist today. I also summarize some of my conclusions so the reader can see the views I have today as a starting point for the journey of how I got there.

Chapter 3-The second Section of chapters cover many of my life experiences, because a reader can't really see how I came to my radical ideas on Reality and the Scientific Method if they don't understand my life's experiences. I had many psychic

On Using the Scientific Method to Study the Paranormal

experiences in my life, but I was also fortunate to go to excellent schools and get an excellent engineering education. These two types of knowledge would seem to be contradictory, and yet this contradiction has given me one of the main driving forces of my life.

Chapter 4- The Scientific Method is reviewed historically and some of it's imperfections are pointed out. The differences between objectivity and subjectivity are also reviewed.

Chapter 5- I discuss how my personal philosophy evolved as a result of my experiences, and how my experiences fit into the modern definitions of sanity.

Chapter 6- I propose some new ideas for how people should rebuild their views of Reality, proposed modifications of the existing Scientific Method, and some ideas for the future.

Chapters 7-8 –These sections discusses some ideas on what the future can hold and scientific advances our future may have by using these new Reality paradigms.

Even though this book discusses a lot of psychic experiences, it is not intended as a book on that subject. There are thousands of books which cover paranormal phenomena in depth. This book uses psychic experiences to illustrate the evolution in my belief system, and to illustrate self limiting views of Reality.

It is also important to mention here that I have never used and do not support the use of drugs to achieve experiences and higher states of consciousness. I have always thought that a person has much more control of their abilities without drugs than with them. I have never used any drug beyond occasional alcohol in any form except for prescribed medications for illnesses.

I hope the ideas I'm outlining in this book can help contribute to new radical breakthroughs to humanity's overall understanding of the universe we live in.

On Using the Scientific Method to Study the Paranormal

Martin Kirk Ettington,

March 18, 2000

On Using the Scientific Method to Study the Paranormal

On Using the Scientific Method to Study the Paranormal

Chapter 1: An Overview of this Book

a. Some Major Dilemmas of Humanity

Here are some of the major dilemmas confronting civilization and science at the beginning of the Third Millennium:

1) Why do two disparate camps of knowledge still exist--that derived from a scientific and experimental view of the world, and the metaphysical/spiritual view of the world?

2) How can Physics achieve a real theory of the Unification of the 4 major forces?

3) How will we ever understand the nature of consciousness in a scientific manner?

4) Can we ever prove or disprove the knowledge of accumulated centuries about metaphysical and spiritual knowledge?

5) What fundamental discoveries of science and reality are left for humanity, and how will we make them starting in the Third Millennium?

6) Will mankind evolve new "psychic" powers and abilities and should he?

7) Can the avenues for self discovery and finding meaning in life be expanded?

I propose to provide a new foundation for approaching answers to these questions from the viewpoint and understanding I've developed in my life.

In this book I hope to give the reader some new ways of looking at Reality and a new orientation for solving problems.

b. My Point of View

On Using the Scientific Method to Study the Paranormal

Since I was a little boy, I always wanted to understand how the world worked and why.

I also endeavored to become a keen observer of our modern civilization with an eye towards improving it and using my God given creative abilities the best I could to help humanity.

My life has generally been a good one. I was raised well by my parents who gave me all the right opportunities.

My career has been a good one although not spectacular, but it does provide interesting work and a comfortable living

I have also been fortunate to have married a loving woman and have wonderful kids.

However, I have always had a core of dissatisfaction with life because I never felt I was really making a real contribution to Humanity and often feel that my talents are under utilized.

One of the guiding themes of my life has been that I always thought I saw something about our civilization and reality that others either weren't interested in or didn't observe.

This observation is very simply that much more exists in reality that our science and technical understanding can account for.

Part of this book discusses how I reached this viewpoint, my logic and evidence for it, and why I think humanity needs to re-orient itself in the new millennium to more fully to understanding the complete reality we live in.

This point of view has also generated an underlying frustration in me with most people's view of the world; because almost everyone I have ever met has lived within limited walls of a self built Reality, which they created for themselves from what they were taught in school and by other people.

On Using the Scientific Method to Study the Paranormal

This is not to say that what people are taught about science and technology isn't true, but that assumptions are made that we know more about Reality than we really do.

The other side of the coin is the mass of uneducated people with no scientific and technical training who accept everything without questioning, and don't try to apply scientific principles to validating their experiences. This non-scientific view is just as bad, because it doesn't lead to a deeper understanding of the how and why of our reality, which at least the scientific method attempts to do.

One way to phrase the question is "Why are people so limited in their views of what reality is?"

I think I've answered this second question as you the reader will see later on in this book, but it has taken me many years.

Supposedly the 1960's started a revolution in raising and understanding consciousness, and that decade did seem to cause an explosion in interest in the psychic and spiritual realms.

What really disappoints me though is that even though many more people believe in the metaphysical side of things today than a few decades ago, there has been very little effort by either the scientifically or metaphysically oriented to integrate the knowledge and experience from both sides together.

This book then is dedicated to providing a clearer model of what reality really is and how humanity can integrate all our knowledge of it together better.

On Using the Scientific Method to Study the Paranormal

On Using the Scientific Method to Study the Paranormal

Chapter 2: A Summary of My Beliefs about Reality

Figure 1-The Concept of Reality

Sometimes it's useful to provide a summary of conclusions to the reader in the beginning. Then the reader can use the rest of the book to see how the logic of these conclusions was reached.

Therefore, I would summarize my ideas on Reality as follows:

1) We live in a much larger Reality than we can measure or observe today, and probably much larger than we will ever be able to measure. This reality can be divided into that which is measurable today, that which we will be able to measure with advancements in our instruments and scientific method, and those parts of Reality which we will never be able to measure.

2) Anything you can imagine exists and has a form of Reality which can be made more concrete through application of the

On Using the Scientific Method to Study the Paranormal

mind. There is no ultimate dividing line between what is reality and what is fantasy.

3) Everything in our physical universe has an extra dimensional energy body counterpart or wave function which can be called a "functional entity". This wave function contains information which determines the properties of the underlying matter or energy. These wave functions can also transmit information to other wave functions through resonance.

4) The power of "Will" is one of the core driving forces of Reality, and Reality is an expression of all of the different "Wills" which created it.

5) The course of future events are determined by a probability which can be controlled by our thoughts. Every event has a probability of occurring, and the magnitude of it's wave function determines how possible that event is.

6) For Science to advance into new realms and subjects, a new enhancement of the traditional Scientific Method is needed which takes into account varying standards of objectivity of measurement.

7) The directions of scientific investigations today like research into a Unified Theory of Physics are missing the integration of the energy body information associated with matter and energy, which will cause many dead ends until this is understood.

8) Science and religion have a lot in common, and they are really not contradictory. The subjective discoveries which are the basis of many religions can enhance the advancement of science as these old subjective understandings can be validated through application of a modified scientific method.

On Using the Scientific Method to Study the Paranormal

Chapter 3: Some Relevant Experiences from my Personal History

The reason I'm including some of my own Psychic experiences is to show that I had an experiential basis to use when considering how to apply scientific reasoning to what happened.

Some persons think that this chapter might be just rambling for my own egotistical purposes. This is not the case. Since I am an engineer and I'm writing about a subject which most people never experience, I think it's very important to understand how I came to validate the reality of the paranormal in my own life . These experiences give me insight to be able to talk about the paranormal from a knowledgeable point of view.

Experiences as A Teenager

a. Ages 11-15 Did ESP Experiments

When I was a young teenager, I had already read numerous books on psychic phenomena and Extrasensory Perception (ESP).

I received a present of a Kreskin's ESP kit, which included a set of cards to test ESP abilities.

I ran an experiment on myself for about six months to see how well I could guess the cards, which would be next in the deck. The results were graphed and I used as many controls as I could.

The results were not promising. I recall that the results did not exceed the expected probabilities and also turned negative at one point to be worse than probability called for.

I did notice that results were best at the beginning and the end of the test, with a big dip in the middle.

On Using the Scientific Method to Study the Paranormal

My conclusions of the possible reasons for the outcome were the following:

1) ESP did not exist
2) I did not have ESP
3) ESP may have other factors such as the emotionalism of the subject, which may affect abilities, and I hadn't tried to measure that.

This experiment made me realize that I was dealing with a subject, which would be very hard to pin down scientifically.

b. First Conscious Out of Body Experience

Another book I read at about age 12 was about "Astral Projection" also known as "Out of Body Experiences" I was really scared, because I thought my mind could leave my body by accident, or something bad could happen to me in the process.

At about the age of 17 I had a very vivid dream where I saw a building in the distance at night, and I seemed to be in a ditch on the side of the road. Then I woke up, but I recalled this place was very real, and it seemed to be a real place.

It was over a year later when I was in college that I recognized that this building was a bar in Albany, New York, which I hadn't ever been to before.

On Using the Scientific Method to Study the Paranormal

c. 1974-1975 Experiments In Kirlian Photography

Another area I did some experiments in was called "Kirlian Photography". This was a very popular subject of study by laymen and some scientists in the 1970s. [W8]

The concept is too use high voltage and high frequency discharges from a Tesla coil or equivalent device to make an imprint of biological objects on 35mmfilm

The belief was that different mental states and emotions could show up as different colors and patterns on the film from changes in the object's electrical field.

I started experiments on my own at home in the summer of 1974 and later worked with some other including my Psychic mentor Sam at RPI in a physics lab.

We made lots of pictures and they were really interesting, but we could never precisely pin down the correlation of mental states to the pictures.

We discovered that there were a huge number of variables, which could affect the results including:

* voltage and frequency
* types of films used
* length of exposure
* moisture on hands
* duplicating the correct emotions with the same intensities each time
* how the film was developed
* pressure on the film

After months of effort my conclusion was that we didn't have the resources or time to get definitive results which would stand up to independent scrutiny by other scientists.

On Using the Scientific Method to Study the Paranormal

Some other work was done at places like the Stanford Research Institute, and they had similar problems.

On Using the Scientific Method to Study the Paranormal

d. My Psychic Mentor

I met Sam Lentine at an RPI lecture in the fall of 1973. He was a fortyish blind graduate student with a Masters in Physics, and who was working on his Doctorate in Physics at RPI.

He was also a very advanced Psychic who not only was well known in upstate New York for psychic healing and doing readings, but he also taught psychic development techniques.

We hit it off immediately and became good friends, and close collaborators on both Psychic Research and
in learning
I helped him organize a class to teach people Psychic development, and he taught me on the side too.

I had read enough to know that psychic development could be done through learning meditation techniques and I was trying to teach myself, but nothing substitutes for having a real teacher.

I learned a lot from Sam, including many psychic techniques such as:

* Learning how to reach a deep state of meditation and relaxation
* Manipulating the "Vital Force" through energy centers in the body
* Psychic healing
* Psychometry (ability to perceive things about a person by an object like a watch of theirs)
* Psychic Shielding

Sam and I also worked together on some research projects like the one on Kirlian Photography

Sam was the first person I had met who not only agreed with my expanded views of Reality, but he had carried these concepts further and had done some investigations psychically too.

On Using the Scientific Method to Study the Paranormal

He was more knowledgeable and experienced than I in the Physics and the Metaphysical was realms, and his validation of my thoughts made me realize I wasn't alone and therefore probably not crazy.

(I only just found out he died in 1991 from medical problems—It was very sad to hear this)

On Using the Scientific Method to Study the Paranormal

e. A Vision which came true

During the summer of 1975 I had a summer CO-OP job at General Electric's Gas Turbine engineering group in Schenectady, NY

At this time I used to meditate at my desk during the lunch hour.

One day in early August I was meditating and thinking about a trip I was planning to Cape Cod. My mind was wandering as I was thinking about what I would do there.

All of a sudden, I had a blinding flash of a scene where I was in the surf at the beach, and a surfboard was coming towards me. Then a shock occurred and I was thrown out of my meditation and was wide-awake.

I thought that this was pretty weird, and mentioned this to a friend or two.

Two weeks later I was walking on the beach on Cape Cod. I saw a couple of guys with surfboards and asked where I could rent one to give it a try.

They said they had an extra one and I could try it with them.

(I had totally forgotten my vision at this point)

I tried to get up on that board all day, and had some modest success, but I was also getting exhausted in the process.

I decided to try it again and fell off when a big wave hit me. Next thing I knew I was coming up to the surface and I saw the exact same scene from my meditation.

The board hit me hard in the chin and almost knocked me out. I staggered to the shore and the two guys I was with helped me to

On Using the Scientific Method to Study the Paranormal

the hospital where they put 10 stitches and 2 sutures into my chin.

The question arises—Would I have been able to avoid the accident if I had remembered my vision and not gone surfing? Later experiences have convinced me that the future is a set of probabilities, and we have free will to decide our actions.

I also had an experience on that trip of being able to partially heal my wounds very quickly through a deep meditation and application of psychic healing techniques. However, I do still have a small scar on my chin from this accident.

On Using the Scientific Method to Study the Paranormal

f. Read a lot more Metaphysics

As these new psychic experiences grew I came to the realization that they were real not in just an intellectual way, but in a strong emotional way which challenged even my admittedly liberal view of reality.

It's the difference between thinking you can fly an airplane and actually doing it. The reality of the experience has a lot more impact than the intellectual thoughts about it.

This caused my state of mind to become somewhat unstable, because I realized more than ever that I needed a new foundation for my belief in reality or "how the world really works".

From my late teens through my mid twenties, I read a wide range of writings from metaphysics, to religion to the occult. The objective was to find or develop a belief system, which took all of the events in my life into account.

One of the clearest books I read on metaphysics was a book written about 1000 years ago called "The Yoga Sutras of Patanjali", which was a scientific exposition on the science of yoga. [B2]

It was originally written in Sanskrit and translated to the English.

It is the only book I've ever read which tries to act as a guide for the mental explorer through the means of subjective experience and understanding of reality as a result of following this guide.

It was very advanced, and although I could see that my experiences related to a few things in the book, the experiences claimed were mainly way beyond anything I had experienced.

On Using the Scientific Method to Study the Paranormal

g. Course on Psychic Phenomena at RPI

In the January Term of courses at RPI in 1975, the administration let students and professors experiment with their own course ideas.

I found out in December that a course would be offered in the Research and Study of Psychic Phenomena by the Physics Department. I went and talked to the sponsor (a Professor Casabella—who was also Chairman of the Physics Department), and told him of my interest, background, and desire to participate in the course.

He agreed, and I ended up as a co-teacher and actually organized and taught most of the course to about 25 students.

The course was research oriented, so we did a variety of experiments to try to document phenomena we had read about.

One of the experiments we did was from what we had read in a book called "The Handbook of Psychic Phenomena", where lots of unusual energies were thought to be focused by Pyramids.

I'd read a couple of books on Pyramids and so I planned an experiment to try different things with Pyramids of different sizes and shapes. These experiments included sharpening razors, growing plants, and several other things.

I have to admit that our results after several weeks with about 50 different pyramids were not compelling evidence of anything – although we thought we saw a few unusual results.
I sent photos and statistics of the results the Rhine ESP Institute Web [W6], and never heard anything—they must have thrown it all away.

This was another learning experience for me on the difficulty of documenting unusual phenomena, but I didn't have enough

On Using the Scientific Method to Study the Paranormal

evidence to say if there were any real effects or I was just not measuring them properly.

On Using the Scientific Method to Study the Paranormal

h. A Frightening Experience with an Invisible Entity

This is the scariest experience of my life. I still can't believe that this happened to me and it's been about 32 years since then.

In the fall of 1976 when I was a softmore at RPI, Sam Lentine was running a psychic development class, which I attended at a private home, in the Albany, NY area.

One of the middle aged women in the class said that she had a haunted house, so Sam decided to investigate it, and three of us went down there one Saturday evening to do this during the month of October. The group included Sam, myself, and a guy named Mark who was also an RPI student.

We also brought devices to try to take measurements, which included:

* A camera with infrared film –to try to take pictures of any entities in the house
* Compasses to detect any magnetic variations in the room
* A tape recorder to try to record any unusual sounds
* Ourselves as sensitizes to try to feel for unusual things
* A two gallon plastic jug of water which Sam said he had psycho-kinetically treated to draw in any entities in the house (His intention was to have the water measured on a Nuclear Magnetic Resonance device at RPI to see if any changes had taken place in the water)

We got to the home in Albany about 9:00PM on a Saturday night, and started taking measurements.
The only unusual thing we could feel were cold spots in the living room which didn't seem to correspond to any vents or air openings in the house.

Sam's perceptions were that there were several entities in the house. I wasn't really sure of any of this.

On Using the Scientific Method to Study the Paranormal

Then Sam took the bottle of water and removed the cap. He placed the bottle on the floor in the living room for about 30 minutes. During this time he said that different energies and entities in the house were being drawn into the water.

Sam then stated that the house we clear and it was time to go.

We went to his friend Caroline's home about 11:30PM and took the bottle in the apartment with us.

Caroline was blind also, and had a Seeing Eye dog, which wouldn't go anywhere near the bottle of water.

Sam and Mark also remarked that they saw fizzing in the water in the bottle. I looked but didn't really see anything.

We ate Pizza and left about 12:30PM or 1:00PM and were driving in my Opel Manta back to RPI on a freeway next to the Hudson River.

I was driving, and Mark and Sam were in the back seat with the bottle of water.

My gas gauge didn't work too well and my car died since it was out of gas.

I flagged down a car to get some gas and was gone about 30 minutes.

When I got back with the gas, Sam and Mark were both frightened and said that they thought something invisible was in the car with them, and that stones had started hitting the side of the car from somewhere.

I still didn't notice anything unusual at this point.

As we started up the road again I really started to feel like something unseen in the backseat was looking over my shoulder and it was really weird.

On Using the Scientific Method to Study the Paranormal

I continued to drive back to RPI, cross a bridge into Troy, and started up the hill, while Sam and Mark become more and more frightened. I said that I thought it was late, we were getting all paranoid, and we should just relax since it was our own fears which were upsetting us.

A minute or two after I said this ---I was psychically attacked. The best way to describe it is that I felt something surrounding my head, and started penetrating into it like an invisible knife. It felt like my brain was being raped.

I yelled and lost control of the car, and started weaving on the road. I managed to gain control again by force of will and pulled over to the side of the street near a storm drain. (It had been raining all evening too.)

Sam and Mark got out of the car yelling that they were being attacked too. We took the bottle, drained the contents into the storm drain, and kicked the bottle into the drain.

Then the attack stopped and we went back to Sam's house. We were so upset none of us could sleep even though it was about 3:00AM by this time. We called a priest Sam knew and asked for a blessing.

For weeks after this I had trouble sleeping because I felt that something was out there waiting to attack me mentally again. Nothing at all like that has ever happened to me before or since that event, but it was real!!—As real as being shot by a bullet. (Which has happened to me too)
If I live to be one hundred, I don't think I will ever encounter anything in the normal course of events—(including a war) which could approach the fear of something attacking my mind like this entity did.
i. Some Meditation Experiences

One meditation technique I learned from reading a book on Zen was to focus your mind on a contradictory statement like "What

On Using the Scientific Method to Study the Paranormal

is the sound of one hand clapping". The idea is that this can lead to a higher state of consciousness. These phrases are called "CO-OMs"

I tried a different one, which is "All the mountains are covered with snow. Why is this one bare?"

After meditating on this for a few weeks, one day I started to feel a "buzzing" and it felt like my perceptions were widening to a volume considerably larger than my body. It was a very altered state of consciousness.
Another time during the summer, I was sitting on the floor in the bedroom a house I was staying in.

I was working on a deep meditative state, and as I did, I started to feel that the objects around me were becoming transparent and I was seeing the reality physics teaches that matter is mostly empty space.

I felt that with a little more understanding, I could mentally change these materials.

A final meditation experience to relate was another one in my dormitory room.

I was trying to get as deep into my soul as I could, and shut out all outside light, sound, and feeling.

(This can be difficult with stereos blaring in a dormitory setting)

I managed to get so deep I didn't perceive any of my surroundings, and I could perceive my soul as a bright light. However, the "I" of me seemed very alone, and I didn't feel any connection to anything else or the underlying reality as some claim.

This may have been a turning point in my search for spiritual meaning, because this aloneness caused me to think that maybe I wasn't touching God, and I needed to try another path.

On Using the Scientific Method to Study the Paranormal

Overall, meditation has given me a variety of experiences to numerous to list, and I recommend it at a minimum as a calming and soothing influence on our stressful daily lives.

I also know from subjective personal experience that meditation is the only tried and true method for most of us to learn and experience our higher natures without danger. (Again, I strongly recommend against drugs for achieving altered states of consciousness.

On Using the Scientific Method to Study the Paranormal

Experiences as An Adult

j. Mental Warning of a Mugging

In 1980 I was moving from Dekalb, Illinois to Rochester, NY between assignments at General Electric, Inc.

While staying with my cousin outside Detroit, I made arrangements one evening to meet an old RPI friend Steve. We decided to go into downtown Detroit to the newly completed Renaissance Center to eat dinner and look around.

The Renaissance Center was built near the water and surrounded by slums.

After dinner we were walking out through the lower level in an area that was all boarded up with nobody else there.

Suddenly, I had this strong urge to turn around and go to find a restroom. I stopped walking forward because the urge was so strong.

I tried to walk forward again and again a very strong urge came to turn around and go back into the main center where other people were hit me.

I remarked to Steve that I couldn't go forward—that something wouldn't let me.

Just then two black guys in trenchcoats appeared about 30 feet away from behind one of the foundation pillars we were about to walk past.

They started walking towards us with smiles pasted on their faces.

On Using the Scientific Method to Study the Paranormal

My friend Steve took off running back into the main area and after a moment or two I figured I didn't know what these guys were carrying under their coats, so I ran too.

In less than 30 seconds we were back in a populated area with Police present, and the two guys chasing us gave us smiles like "next time w'ell get you" and took off going the other way.

I had previously always tried to pray to God for protection, and tried to give a subconscious message to my senses to warn me of danger.

I'm convinced that whatever sense or "angel" warned me that evening, I would have been killed or severely wounded if I had continued walking out of the complex with no warning.

On Using the Scientific Method to Study the Paranormal

k. Warning of a Disaster

During early August of 1998, my wife and I decided to send her and our kids to visit her mother in Barcelona, Spain.

I was going to buy a ticket separately, and meet them there during early September.

When I called the travel agent to book my ticket I had a terrible feeling of fear about taking the flight.

I tried two other times to book the ticket during the week for a September 2nd departure, and each time I got the same strong feelings of fear and possible death.

I have always prayed and tried to guard myself mentally to avoid disasters, so finally I took the warning seriously and decided not to go at all.

This was really very difficult to do since I really wanted to see my wife and kids, and this meant I would be home alone for a month.

Work wasn't an excuse either, since I wasn't doing any really heavy contract work at the time and could easily have taken the time off.

I called my wife and told her my decision, and she was surprised, but agreed for me to follow my instincts.

On September 2nd the Swissair disaster occurred on a plane leaving Kennedy airport in New York, which crashed in Newfoundland Canada with all lives lost.

I would not have originally been booked on that flight, but could have easily ended up on it since I was due to fly through Kennedy airport, and any delay might have caused me to switch planes

On Using the Scientific Method to Study the Paranormal

On Using the Scientific Method to Study the Paranormal

Chapter 4: THE SCIENTIFIC METHOD

a. What is the Scientific Method?

It is always a good idea to periodically evaluate one's progress personally in life, and in any field of endeavor. In science, we should also occasionally re-evaluate our models of objectivity to see how science is proceeding, and it may be that scientific objectivity overall needs to be re-examined.

Although many of the readers of this text may be well educated technically, I know that most technical schools don't specifically teach the history of the scientific method, so a review of the scientific method is called for to begin my case of how it should be updated.

This Scientific Method became informally popular during the Renaissance when scientists like Copernicus used his observation of the planets, and the fact that the earth was not the center of the cosmos, to show that existing theories of the cosmos were wrong.

Since then, the entire fabric of our civilization has been based on theory and experimentation.

Science is firmly based on hypothesis and theories, which can be objectively proven or disproven through repeatable experiments.

The history of the Scientific Method was also heavily impacted by three historical persons, who were also giants in philosophy, mathematics, and science in the last Millennium:

1) St. Thomas Aquinas
Saint Thomas Aquinas [B6] lived from 1225-1274 and was a Catholic priest who contributed greatly to the discussion going on at that time about Science Versus Religion.

On Using the Scientific Method to Study the Paranormal

A quote from a website on St. Thomas Aquinas [W4] sums up his effect upon western Christian faith and reason:

Faith and Reason. -- The principles of St. Thomas on the relations between faith and reason were solemnly proclaimed in the Vatican Council The second, third, and fourth chapters of the Constitution "Dei Filius" read like pages taken from the works of the Angelic Doctor. First, reason alone is not sufficient to guide men: they need Revelation; we must carefully distinguish the truths known by reason from higher truths (mysteries) known by Revelation. Secondly, reason and Revelation, though distinct, are not opposed to each other. Thirdly, faith preserves reason from error; reason should do service in the cause of faith. Fourthly, this service is rendered in three ways: (a) reason should prepare the minds of men to receive the Faith by proving the truths which faith presupposes (praeambula fidei); (b) reason should explain and develop the truths of Faith and should propose them in scientific form; (c) reason should defend the truths revealed by Almighty God. This is a development of St. Augustine's famous saying (De Trin., XIV, c. i), that the right use of reason is "that by which the most wholesome faith is begotten . . . is nourished, defended, and made strong"

2) Rene Descartes

Rene Descartes [B7] lived from 1596-1650 and was known as one of the greatest philosophers of all time and was also the inventor of analytical geometry. His famous quote in Latin of "Cognito Ergo Sum" which translates to "I think therefore I am" was the result of a lot of his meditations on the nature of consciousness.

One of his most famous philosophical treatises on the scientific method is titled "Discourse on the Method of Rightly Conducting the Reason, and Seeking Truth in the Sciences", and it expounds on many aspects of consciousness including deciding what was truth and what wasn't.

In Part 6 of this treatise he remarks on the value of experimentation versus use of just the senses:

On Using the Scientific Method to Study the Paranormal

"I remarked, moreover, with respect to experiments, that they become always more necessary the more one is advanced in knowledge; for, at the commencement, it is better to make use only of what is spontaneously presented to our senses, and of which we cannot remain ignorant, provided we bestow on it any reflection, however slight, than to concern ourselves about more uncommon and recondite phenomena: the reason of which is, that the more uncommon often only mislead us so long as the causes of the more ordinary are still unknown; and the circumstances upon which they depend are almost always so special and minute as to be highly difficult to detect."

3) Sir Francis Bacon

Sir Francis Bacon [B8] who lived from 1521-1626 is largely credited as the father of the modern scientific method we use today.

A description of him by Voltaire (another famous French philosopher) credited Sir Francis Bacon as the person who made "experimental philosophy" a common feature of science in Europe.

A quote about Sir Bacon's work on experimental philosophy follows:

"In a word, no one before the Lord Bacon was acquainted with experimental philosophy, nor with the several physical experiments which have been made since his time. Scarce one of them but is hinted at in his work, and he himself had made several. He made a kind of pneumatic engine, by which he guessed the elasticity of the air. He approached, on all sides as it were, to the discovery of its weight, and had very near attained it, but some time after Torricelli seized upon his truth. In a little time experimental philosophy began to be cultivated on a sudden in most parts of Europe. It was a hidden treasure which the Lord Bacon had some notion of, and which all the philosophers, encouraged by his promises, endeavored to dig up."

On Using the Scientific Method to Study the Paranormal

The civilization we live in today has been largely created by application of the scientific method, which was in large part formalized by the above three persons.

Our understanding of Science and Natural laws of reality allows engineers and others to create the things, which make our life easier. This includes everything from electricity to cars to spacecraft.

The Scientific Method can be expressed this way:

1) Develop a hypothesis of how something works

2) Make an experiment to test the hypothesis

3) Modify the Theory resulting from the hypothesis to conform to the results

4) Keep testing until the theory is proved or disproved.

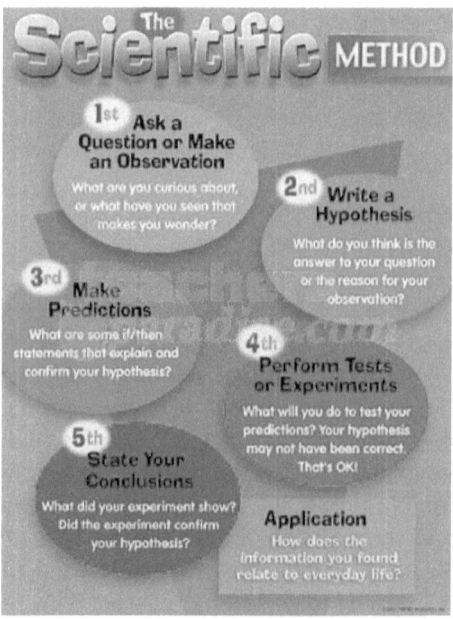

On Using the Scientific Method to Study the Paranormal

Figure 2-The Process of the Scientific Method

b. The Non Scientific Method of Observation

This has been the dominant way that man observed the world for most of his history.

Man would observe something then develop a theory about it. There was no requirement that it was necessary to prove the theory to be correct or incorrect.

Thus some major misunderstandings about Reality were "Set in stone" such as the concept from Ptolmy in ancient Egypt, that the earth was the center of the cosmos. This flawed theory of the cosmos dominated for centuries and was probably responsible for holding back the true understanding in many areas of science.

Although the Roman Catholic Church has done many good works in history, it is also responsible for the stagnation of many areas of Science by holding church doctrine on the truth to be more important than real observations of the truth. In fact it wasn't until the 1990's—almost 500 years after the fact, that the Pope issued paper stating the Copernicus was correct after all.

On Using the Scientific Method to Study the Paranormal

c. Science Versus Pseudoscience

It's critical to the understanding of my approach to a new philosophy to outline the differences between the scientific and pseudoscientific points of view.

Modern science is the foundation of our civilization and it's methods need to be fully understood as well as it's limitations to allow the reader to see where my thinking is proceeding.

Although in many parts of this book I criticize the Scientific Method , it has been critical to building the foundation of our civilization today.

Most people in the world have never been taught the Scientific Method so they don't questions events they observe the way they should.

Many of these non Scientific Method people accept claims of events without the proper scientific scrutiny. This is a Pseudo Scientific attitude which is too accepting of new phenomena and doesn't advance the scientific understanding of reality.

Most of human history has been written by people with this non scientific attitude, and although civilization advanced slowly by trial and error,(such as the discovery of fire and farming); the lack of understanding of the fundamentals of nature and the lack of emphasis on repeating results limited advancement and caused false understandings of the ways things worked to continue for centuries.

Ignorance has led to many beliefs about God and reality for thousands of years which may have been wrong or incomplete.

One traditional example is the ancient's beliefs in the four fundamental parts of matter of Fire, Water, Earth, and Air. This sounds like a good division of matter, but has no underlying experimental observations to support these divisions.

On Using the Scientific Method to Study the Paranormal

Once people started really observing and recording Reality, they made progress.

The beginning of the use of repeatable observations was really the beginning of the scientific method.

The methods used to observe were primitive and depended mostly on the senses of the observer. Primitive could be effective though if the results were repeatable.

Many times though, people made a hypothesis based on their beliefs and didn't think it was necessary to test them for validity. This became pseudo science and stunted advancement.

An example of Pseudo Science was that most people thought that life arose spontaneously from non life until Evolution through Natural Selection became accepted through the work of Charles Darwin.

On Using the Scientific Method to Study the Paranormal

d. Subjective Versus Objective Knowledge

The scientific method is by it's nature objective. In other words someone can independently verify all scientific observations with the right equipment and procedures.

This objective knowledge provides a strong foundation for building other science knowledge and technology, which makes our daily lives easier.

Subjective knowledge on the other hand is knowledge, which only the observer has. It has not been verified independently.

A good example of subjective knowledge is someone who has a strong vision of Jesus. The observer may feel that it was a real experience, but they usually can't prove it to someone through objective experimental results because only the observer was aware of the vision.

e. Can Everything be measured Objectively?

Today we have so much confidence in the results of technology from application of the scientific method that hard scientific advocates assume that everything can be measured by experiment and proven true or false.

The logical extension of this reasoning for most people is that if something can't be measured, it doesn't exist.

Now of course many people believe in God, and they know God can't be measured. Even many scientists profess belief in God.

This belief is not logical if God can't be measured. Right?

However, scientists also admit that there are many clearly physical phenomena which they can't measure either.

Examples might include:

On Using the Scientific Method to Study the Paranormal

1) What is going on at the center of the earth?
2) Are the fundamental constants of the Universe the same at a point several light years from us?
3) What happened to the Universe in the first instant of the Big Bang?

These are questions which may never be answered with objective scientific observation

This limitation is basically one of instrumentation. If you don't have an instrument to measure something you can't experiment on it to generate objective results.

An example would be that there was no way to measure radiation when the understanding of radiation was too limited to have already developed instruments to measure it.

This is an old problem which is part of what science is all about. Instruments have to be developed to make observations with the proper accuracy to prove or disprove a hypothesis.

There is a further problem with measurement:

Are there phenomena which exist in the world of consciousness which we don't haven't instruments to measure?

The anecdotal evidence is that the answer is yes—many people think they have experienced certain phenomena, (like telepathy) but we don't have any standard objective instruments to measure if it exists or not.

Here is another question to think about:

Are there events of consciousness which we can never develop traditional instruments for because they exist outside of our physical world?

On Using the Scientific Method to Study the Paranormal

The answer to this line of reasoning now becomes clearer—that there certainly are phenomena which we don't presently have the instruments today to properly measure. (How do you measure Love or Hate in a person?)

Even more discomforting than this is that there may be phenomena which we will never be able to measure even though we have a pretty good subjective idea that they may exist.

(How do you measure a ghost or a vision of the future—or the first instant of the Big Bang which created the Universe?)

We have become very confident late in the 20th Century that we are close to understanding reality and close to developing an integrated theory of all the physical forces.

We think we are close to final answers in the sciences and in shutting the door on ignorance and phenomena, which we can't verify objectively.

It's ironic that some scientists at the end of the 19th century also thought that they had discovered everything and there was nothing new to learn—just like the general feeling at the end of the 20th century.

Maybe the problem is that our vision of reality as proven by the scientific method is too narrow, and until we expand the scope of our thought to include a way of understanding other phenomena not so easily measured; we will never really get to a breakthrough understanding of reality.

We should also all have less hubris and more humility in our estimation of how far science has come and how far it has to go.

On Using the Scientific Method to Study the Paranormal

Chapter 5: MY PHILOSOPHICAL EVOLUTION

a. Are Psychics Mentally Ill?

Mental Illness has often been considered a reasonable scientific diagnosis for people who have experienced the types of things I have in my life.

If the time was several hundred years ago or earlier I would have a good chance of being executed or burned at the stake for talking about some of these experiences

Fortunately, we live in a more liberal thinking time when most people with mental illness are put into institutions to take care of them better.

Even more fortunately, (at least in this country), we are allowed to express weird views and in the last few decades not be institutionalized unless we were a danger to ourselves or others.

I don't think my experiences and ideas are a danger to society-- but who can tell.

Seriously, I have met many people who claimed to have had psychic events in their lives. Most were pretty stable individuals, but a few had spent part of their lives in mental institutions—or belonged in one.

What is considered normal behavior by a witch doctor in one society is considered mental illness in another.

I wonder how many people in institutions today may be having valid experiences because they are very sensitive, but the experiences have overwhelmed their mental stability and ability to function in society on a daily basis.

On Using the Scientific Method to Study the Paranormal

I have read certain quotes such as "Genius is Close to Insanity" and I've tried to picture before how one's consciousness can move from one state to another. I think I can see the line.

I also remember a psychology class I once took at RPI where a movie was shown of mentally ill persons.

One person in the movie said that "his thoughts were turning into insects". This is very obviously a strange statement which seems to have no logic discernable. On the other hand, if you evaluate the statement in a meditative state from another point of view, the statement is no less strange than the famous Zen statement of "What is the sound of one hand clapping".

Another question is my self-diagnosis of my own mental state. This is of course a totally subjective evaluation.

However, based on the fact that I know I have an extremely stable personality, and have always been able to cope with life better than most, I don't think I have anything to worry about.

In fact, I submit to the reader that being able to ask this question of one's self, and one's experiences says a lot about the intelligence and stability of the person asking the question.

Finally, my own thinking says that my psychic experiences were valid for my then state of consciousness, and that using Occum's Razor, the easiest explanation for what I've observed is that what I've observed is real. The rules of our civilization and ways of thinking just ignore many of these experiences, because no scientific framework exists to explain them. (Occum's Razor is an old saw of logic which says that the simplest explanation is usually the best one to explain an observation)

On Using the Scientific Method to Study the Paranormal

b. Why do we search for the meaning of life?

Since time immemorial, people have wanted to understand their environment.

Initially this search was probably fueled by a desire to understand the overwhelming forces of nature which controlled the lives of primitive man. These included fires, floods, volcanoes, earthquakes, etc.

Therefore religion was developed and man made his best efforts to connect with some type of God or Gods.

Thousands upon thousands of books have been written, and major religions founded on the question of "What is the meaning of life?", so I will not try to re-address that issue here except to note that there is a burning desire for people to find meaning in their lives.

Throughout history, people have turned to religion and spirituality to find meaning in their lives.

Part of this question of the meaning of life has been transmuted in modern times to a more scientific investigations of our Universe, and to use methods which are more objective than what primitive man used which was basically his God given senses.

However, today we live in a civilization run on scientific principles which seems to be totally at odds with a religious/spiritual view of the world, and many people try to find satisfaction in the physical fruits of the resulting production of our civilization.

Our civilization provides physical abundance unprecedented from earlier centuries, but people seem to be even more unsatisfied with their lives than in the past.

On Using the Scientific Method to Study the Paranormal

According to the modern scientific view you would think people could find meaning in their lives though the modern discoveries of science and technology which provides more comfort and leisure.

Why then do so many people (including this writer) feel a need to follow a certain religion even though they are well grounded in the scientific disciplines and scientific/technical mode of thinking?

This is because there is a natural progression of people satisfying their basic needs of food, shelter, and clothing, etc. so they are looking for higher needs to satisfy, and religion often satisfies the Self Actualization need.
(Maslow's hierarchy) [B5]

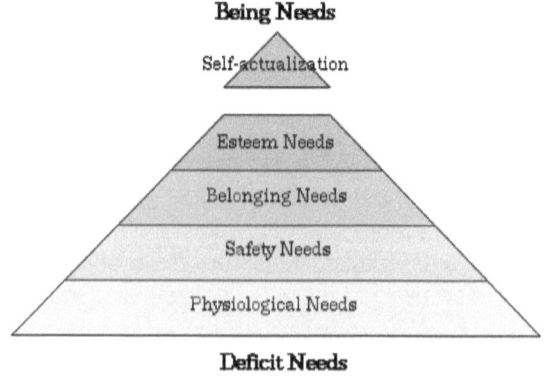

Figure 3-Maslow's Hierarchy of Needs

I believe we are witnessing a serious and growing disconnect between the scientific and spiritual points of view.
These two camps of thought don't want to give ground to the other.

Scientists think that they will eventually be able to get people to give up their superstitions and unjustified beliefs for a "modern

On Using the Scientific Method to Study the Paranormal

view" of reality, while the metaphysical/religious side believes in things which are unproved today and may always be so.

I believe both sides are right and wrong to some extent and a new view of reality is needed.

Hopefully, this will provide more acceptable avenues for people to explore to meet their full potential and find meaning in their lives.

On Using the Scientific Method to Study the Paranormal

c. How my thinking has evolved

You should understand that I was still a pretty conventional thinker about Reality up until some of the vivid experiences I had in my late teens and early twenties.

When I first started to have solid experiences like being able to manipulate life force energies in my chakras [B10] , and being able to really do "mind reading" it was a real shock to my mental stability. This was since these concepts went from my understanding them in an abstract way to real experiences, which I knew were real but not accepted by our civilization.

I actually felt I was "starting to lose it" a couple of times, because my experiences were scary enough to cause me to question my own sanity.

On deeper reflection however, I decided that yes--these experiences had occurred, and if the results were bothering me because they were undercutting the foundation of the Reality that I had built, then my understanding of Reality must be wrong and that was what needed revision.

The next years of my life were in large part dedicated to developing a better framework of Reality, which really conformed to the observed facts, which had confronted me.

On Using the Scientific Method to Study the Paranormal

Chapter 6: TOWARDS A BETTER FRAMEWORK OF REALITY

a. Our Limited Understanding of Reality

Our scientific understanding of the world is only about 500 years old.

Mankind has existed for over 100,000 years, and the universe is billions of years old.

Is humanity so arrogant as to say that we have a close to final understanding of the natural scientific laws of the universe, or should we be more humble and admit that we only understand a tiny fraction of what is out there, and much more is undiscovered than discovered.

I spent many years reading articles and journals from organizations like the ASPR (American Society for Psychical Research) [W5] who have done good experimental work for 60 years on validating and understanding psychic phenomenon.

However if I were to go to the average person on the street they would say that these things have never been proven.

(I also read the standard scientific journals like Science and Scientific American.) Most scientists would also say that paranormal events haven't been proven to exist.

Instead of exploring how to understand and benefit from these abilities, most researchers in these areas are still being asked to prove that these things really exist. In this case many of the skeptics aren't really interested in the objective evidence because it would disrupt their cozy worlds.

On Using the Scientific Method to Study the Paranormal

b. The Scope of Reality

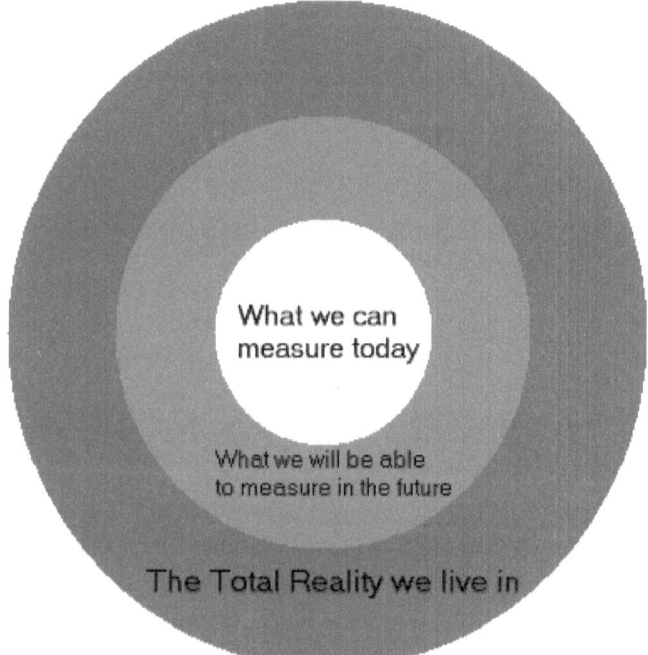

Figure 4-The Measurement of Reality

From the foregoing discussion on the scientific method and what is measurable, you can tell that I must have a significantly different idea of reality than the norm.
Diagram #1 best illustrates my belief of our ability to understand Reality:

1) The inner yellow circle represents what we can measure with our instruments today and perform experiments on to prove or disprove theories.

2) The red circle is a larger area, which we will eventually be able to measure to understand and prove or disprove the way things are

On Using the Scientific Method to Study the Paranormal

3) The blue outer circle is the largest area, and is that part of the universe which we may be able to experience but will never be able to measure and validate with objective scientific approaches.

We may be able to subjectively perceive a lot of things in the blue area, but will never have the tools and techniques to objectively quantify it.

This blue realm may also include such things as where the soul goes after death, the fundamental nature of God, and certain dimensions of space and time, which we can postulate but never prove or disprove.

The red region may be more amenable to creative approaches for objective measurement and validation.

However, there will have to be agreement among the scientific community on some new approaches, which may constitute legitimate standards for objective measurement of phenomena.

This may include indirect evidence, which is used in areas like particle physics.

Neutrinos for example can't be directly perceived, but their existence can be inferred by collisions with other particles, which make cloud tracks which we can directly perceive.

The same approach should be transferable to validation of something like telepathy, where the medium of thought transference may not be understood at this point, but it can be validated through well-controlled blind studies and statistics.

I think that this type of validation issue of the objectivity of an experiment also presents barrier to further scientific progress.

On Using the Scientific Method to Study the Paranormal

Until new objectivity standards are set, we will never make good progress on a scientific understanding of consciousness and "non physical" phenomenon.

On Using the Scientific Method to Study the Paranormal

c. The Scale of Believability

When I was in my second year as an undergraduate engineering student, I had an opportunity to run a psychic research course during the 1975 January term at RPI under the auspices of the Chairman of the Physics department.
(He scheduled the course and I worked with him to develop the material and taught a lot of it.)

At that point I had also read a lot of books on metaphysics, and was a year into my own psychic development through meditation.

I was the daily instructor to a class of about 25 other students, where we did experiments on everything from pyramids to trying to hear messages from the dead on recording tapes.

In looking for a way to explain psychic phenomena to these students I came up with my scale of believability which I have become an ever-stronger adherent to as I've gotten older.

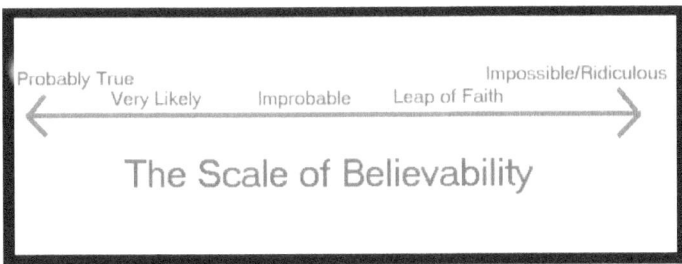

Figure 5-The Scale of Believability

The concept is simple:

Imagine a scale with one end called "Probably True", and the other end called "Impossible/Ridiculous"

Then start listing unusual types of phenomena along the scale as to where they fall.

On Using the Scientific Method to Study the Paranormal

An example scale might include the following key, with these sample entries:

Probably True
Very Likely
Improbable
Leap of Faith
Impossible/Ridiculous

Telepathy-Probably True

A very common occurrence is when you think of someone and they call on the phone right away; or how many married couples think the same thoughts at the same time.

Most people would agree because of these events, telepathy probably exists as a natural human ability in some form although it hasn't been validated by the accepted scientific standards yet

Psychic Healing-Very Likely

Many persons throughout history have reported natural healing from Christ to people in today's world. A lot of anecdotal evidence exists, and this is a common enough claim that most people would accept it's existence, but they are still somewhat skeptical. Of course the scientific community doesn't validate this either.

Foretelling the Future-Improbable

Again, something commonly claimed. The standard story being someone who dreamed they saw a plane crash and then it happened. Now we are getting to something that is very subjective and can't be validated by today's scientific techniques.

Out of Body Experience-Leap Of Faith

On Using the Scientific Method to Study the Paranormal

Many books are written on this subject, and thousands of cases have been reported, but it is a totally subjective experience and not measurable objectively today

Mental Teleportation-Impossible/Ridiculous

I'm talking about Star Trek beaming capability without the machines. This is really getting to the edge, but again if you search the literature, there are cases of this event reported throughout history although it is extremely rare.

Many more items can be added to this scale. I just wanted to use some sample ideas to explain the concept.

On Using the Scientific Method to Study the Paranormal

d. The "Real Reality"

After I made the preceding scale on Believability to organize some of the paranormal phenomena I had learned about,(and experienced some of) my next inclination was to try to create a dividing line as to what could be real and what was fantasy.

I found out I couldn't do it. Here is where I left the objective frame of reference and started to depend on subjective experiences, my logic, and my intuition:

I came to the pretty far out conclusion that not only does everything exist, but in the total of the "Real Reality" everything does occur, everything can happen, and everything probably has happened somewhere and sometime.

(The above statement has some similarities to the Buddhist and Hindu views of Reality which say that it is all an illusion. My statement also has some similarities to what science knows about matter which is that individual atoms take up very little space, and so the objects we see around us are really mostly empty space.)

In other words we live in a tiny self-restricted reality—anything you can think of is possible in some state, and ultimately there is no fantasy. Anything you can think of has some state of reality.

Does this mean the average person can think of a pink elephant and make it appear? Its Unlikely—but all things are possible given the right circumstances.

I know that this concept sounds insane, but it is the only view of reality which allows the explorer to think outside the traditional reality box and set their own boundaries on what they can observe and do –not someone else's.
This concept is anathema to the scientific method because science is all about defining the differences between reality and

On Using the Scientific Method to Study the Paranormal

what is not reality. If everything exists how can science determine what is really true?

Well, science has a wonderful purpose in giving us rules to let us mold reality, but most people get carried away into thinking that if it hasn't been proven it doesn't exist, which is a logical fallacy.

Can I prove my theory of Reality--No--but on the other hand many theories can never be fully objectively proven at all.

On Using the Scientific Method to Study the Paranormal

e. Fantasy and Reality are obviously different

There is an obvious flaw which most people would immediately perceive about my statements on their being no difference between fantasy and reality.

It is the simple observation that a cartoon dog on the television (fantasy) doesn't ever have the same physical impact as a real dog (physical) which you can pet and play with.

How can this obvious difference be rectified with the fact that are observations of these two things are different?

I propose that this difference is a merely a difference in the frequency and resonance levels of the two, and that the observer could tune into the cartoon reality through putting themselves in the correct state of consciousness.

On Using the Scientific Method to Study the Paranormal

f. On the Possible nature of Consciousness and Reality

One of my close friends and associates from my RPI days was Samuel Lentine, He was the blind Physics graduate student I've previously mentioned.

Sam went on to get his PHD in biophysics at RPI, and had at least three masters overall—in education, natural science, and physics as an educational basis for doing his experiments.

He became an instructor at RPI and did work in the BioPhysics department as well as starting a holistic clinic on the applications of psychic healing and psychotronics to everyday health.

Sam had a theory about the nature of Reality which ties in well to my own observations. I am summarizing his theory mainly from what I've listened to on taped lectures he gave during the 1980s to an organization called the US Psychotronics Association.

His theory was based on the idea of what he calls a "wave function" or "functional entity".

The basic premise of his theory was that all matter (and even space and time) has a wave form associated with it which is really a form of consciousness, and can be described as an extra dimensional attribute of matter.

You can call this energy the vital force, or orgone, or psychic energy, or consciousness, or many other names, but it is basically an energy field of the vital force of consciousness which is associated with every piece of reality down to the sub atomic level.

This belief is also in line with the Yogic and Buddhist religions which believe that all parts of the physical reality are conscious at some level.

On Using the Scientific Method to Study the Paranormal

Sam believed that this energy field contained an information imprint which described the nature and properties of every object, and that the physical object's properties would conform to the information which was imprinted on the energy field.

He also believed that this energy field could be described as a wave function and that it could be modified by other wave functions communicating with it and imprinting their waves onto the object's wave function.

One experiment he conducted was to imprint the wave function of the element radium onto water and measure the difference in the radiation count resulting. He claimed significant results from this experiment.

The experiment was done using psychotronic devices and mental manipulation to do the imprinting, and standard scientific instruments to measure and record the resulting radiation counts.

This type of experiment was performed using a hybrid of standard scientific instruments and psychic techniques which I have recommended as needed to expand the range of objective measurements needed to properly observe some of these phenomena under scientific conditions.

This theory would explain many of the paranormal phenomena which have been described, since many of them involve one object affecting another through paranormal means.

An example of how this works would be telepathy. Telepathy may work when the energy field or wave function of one individual communicates information which imprints the wave function of another individual. This imprint communicates the thought information to the second person which this other person understands.

How is this imprinting done? Sam and I both believe it done with some type of resonance between similar wave functions.

On Using the Scientific Method to Study the Paranormal

Resonance occurs when two waves of similar frequency interfere with each other. The interference can enhance the target wave or cancel out the target wave depending on the amplitudes and structures of each wave function.

An example of resonance is the use of a tuning fork. The first tuning fork will cause the second tuning fork of the same design to vibrate when the first tuning fork is struck and the sounds waves transfer to the second one.

Sam's theory ties well into my own beliefs on Reality especially in what I describe as the "Scale of Believability".

If it is true that all physical reality is the result of the imprinting of information from conscious wave functions, then the idea that whatever a person believes can become reality makes sense.

This means that whatever a person thinks modifies the information content of their own personality wave function, and that the proper communication of this information to another person's wave function enables telepathy.

This also has impacts on the way physics, biology, and other sciences measure things because it means they are leaving out a significant component of reality from their measurements.

This view of Reality also does not contradict my views of God's relationship with man as Christianity propounds, it merely proposes a mechanism as to how this interaction occurs.
g.   What is the Force of Will?

One more belief I have is in the power of a person's Will. What is Will Power? The dictionary defines it in part as--used to express determination, insistence, persistence, or willfulness OR mental powers manifested as wishing, choosing, desiring, or intending

A person's Will seems to be a central component of their personality. This Will also seems to be the driving force behind accomplishing anything in our Reality.

On Using the Scientific Method to Study the Paranormal

I think the human Will is also the driving force behind any of the phenomena discussed in this book.

People can develop their Will through exercise and experience like any other ability. (Have you ever noticed how two different people can say the same thing to a third person with different results. One reason for this is the will behind the person try to exert their influence.)

This Will can also be used to direct all of the actions both Psychic and non Psychic which are possible.

Maybe one way of looking at Reality is that the individual's Will acts on others Will, and on the Will which created the Universe (God?) as a whole. (Is it blasphemous to say this? I don't think so, because even the Bible says that God gave man free will, and man obviously can affect Reality which is God's overall creation. Therefore this is part of God's plan.)

The resulting Reality is a combination of the effect of all of these forces of Will.

We could also say that the Will is the uniquely identifying characteristic of a real Intelligence. Computers can analyze, but they don't have a Will.

On the other hand, even a worm has some type of will to decide what direction is bores holes in.

Will is the guiding influence which directs many forms of energy to accomplish or create things.

A person can therefore change the impact of themselves upon Reality by changing the force of their Will.

On Using the Scientific Method to Study the Paranormal

h. The Probability of the Future

One of the things I learned in my study of Reality was that everything has a probability of happening, and we can control those probabilities through our thoughts as well as our physical actions.

A physical action which could affect my future, would be where if I were to study in college and get a degree in Physics, I would increase the probability that I could become a PHD Physicist.

However, I'm not talking about just what can done through physical acts to affect one's own future. It is also possible to change the probability of an event happening through application of energy and information to the wave function for the probability event.

An example was one time when I was asked to join a football betting pool of about twenty teams and how they would do on the coming weekend. I didn't know anything about the teams, so I just mentally looked at the probability of which teams would be likely to win in a contest. I won the pool since my guesses were more correct than anyone else's guesses—even though the majority were very knowledgeable about the teams in question.

Another event which I've always questioned was my story about how I had a vision of getting hit by the surfboard, and then it later came true.

If I had made a real effort, I should have been able to avoid the getting on a surfboard in the first place and therefore avoiding the accident entirely.

Even religions like Christianity say that people have free will, and we are masters of our own fate.

On Using the Scientific Method to Study the Paranormal

I look at the future like we are in a river, going with the current (the time direction), and we have free will to navigate between the banks.

We also have an ability to look downstream and see rocks or shallow areas to avoid in the river. We can change direction to avoid problems depending on how well we look ahead, and take advice from our physical and non physical senses.

I had two other events which demonstrated to me the ability of a person to change their fate:

- When I was in the Renaissance center in Detroit in 1981, and had the mental warning of a mugging, I had time to stop and change direction before the attack occurred.

- I also had a warning of a possible disaster happening to me if I took a certain plane flight to Europe as I discussed in the chapter titled "Warning of a Disaster". I did not take the flight which crashed and am here to write this down as a result.

The probability of events occurring is another wonderful aspect of the Reality we live in.

On Using the Scientific Method to Study the Paranormal

Chapter 7: WHAT THE FUTURE CAN HOLD

a. The confrontation of Science and Spirituality

As previously discussed, there are two main camps of belief in Reality in the world today. These two groups I will call the "Scientific View", and the "Spiritual View".

Some people have feet in both camps, but most people belong entirely to one or the other.

There is a confrontation between the two groups, which can be characterized with the following imaginary points of view:

The Scientist's Orientation—

I believe in what can be proven by experimental evidence. Science has taken civilization out of the dark ages by grounding us firmly in Reality.

The spiritual forces like the Roman Catholic Church fought hard against early scientists like Copernicus who challenged the church's view of the way the heavens worked, and they were proven wrong.

If scientists hadn't challenged assumptions like these and others about reality, we would all still be living in primitive ways.

Religion and Spirituality are therefore dangers to scientific advancement, and people should be educated to the modern worldview, which has provided a standard of living unmatched at any time in the past.

I know from scientific evidence that what science can do to manipulate reality is real and true.

People who are ignorant tend to be Spiritually oriented because they don't know anything better.

On Using the Scientific Method to Study the Paranormal

The Spiritual Orientation—

There are many questions people ask about who created us, and the purpose of life, which can't be answered by Science at all.

Science for all its wisdom doesn't have any better idea about the fundamental nature of consciousness that it did 100 years ago.

There are many experiences which people have had over the centuries, which can't be explained by Science; except to deny that the experiences ever happened. Is that really objective?

In fact, I know from personal experience and faith in my religion that it is real and true.

Scientists may have done a lot to improve our physical well being, but is that an end in itself?

Is the ultimate aim of science for us to understand and control nature without any comprehension of the larger questions about God and our purpose on earth? This would not be a good way to live our lives.

a. A Simple Neutral Analysis:

* Both viewpoints have logic to them.
* Both views have good arguments that the other one could be a danger to our civilization if unchecked
* Both views seem to be diametrically opposed.

b. A Deeper Neutral Analysis:

The scientific viewpoint has only come into being in the last 500 years. Before that only the spiritual viewpoint had a real influence on the world.

On Using the Scientific Method to Study the Paranormal

Now that science is maturing, and is the driving force in the growth of civilization, it may have become arrogant in assuming that the traditional pure scientific view has an answer to everything.

Both views have elements of truth in them, but we are just at a point in our maturity as a race where we may be wise enough to integrate the best elements of both worldviews together.

On Using the Scientific Method to Study the Paranormal

c. A Synthesis of Science and Spirituality?

Some people feel comfortable with an integration of their scientific understanding of the world and their spiritual views or religion. St. Thomas Aquinas [B6] was one of the first great thinkers in history to work on a merging of the two, but most people still see them as opposing points of view.

How can we get the best merging of the objective information provided by Science, and the subjective information from the Spiritual point of view?

Here I would like the reader to see the benefits of such a collaboration:

What if people would realize that the quest for truth would only be strengthened by using both the Spiritual observations and Scientific methods?

Example #1—

Experiment:

A study could be done to measure the effects of spiritual healing on people using controls and procedures which most scientists could accept. The measurements should include both traditional scientific instruments and subjective evaluations from the patients.

Results:

If the results were positive and accepted, this would both expand the boundaries which science can explore to understand the energies of life, and provide evidence for more people to accept the benefits of that particular spiritual point of view

Example #2—

On Using the Scientific Method to Study the Paranormal

Experiment:

Arrange a multiple participant study of Astral Projection(OBE). Decide on acceptable success criteria beforehand. Get agreement from scientific community that if successful they would agree the results are valid. Use common subjective validation of physical locations for provide objective verification.
Results:

If the results were successful this would provide another insight into the understanding of consciousness, and lead more people to try to experience OBE in their spiritual quests. It might also provide new avenues for Physics to evaluate that they might be some missing some forces of the Universe in current models.

Example #3—Here is a long-range idea to explore:

The mind can imagine itself at far places that the body can't presently go to like another star.
Some proponents of Astral Projection claim they can travel spiritually to another planet or location as fast as thought.

Results:

Would validation of this capability eventually provide some new modes of extra dimensional travel to other stars for the human race?

Overall, I hope you can see that there are many potential benefits to our daily lives from people widening their narrow views of Reality to encompass the possibility of the validity of other viewpoints.

When we close our mind to what may be real, all we do is reduce the possibilities of what can be explored.

On Using the Scientific Method to Study the Paranormal

d. A Roadmap to a New Way of Thinking

We come into this world as babies without understanding anything about what is going on.

As we grow we learn through our senses what is going on around us, and start to formulate our views of the world.

This foundation for understanding Reality is built from birth and heavily influenced not only by what we observe, but also by what we learn from those around us.

A person who grows up in a tribe in the Amazon jungle probably takes it for granted that there are spirits in the trees, and that their ancestors are watching over them.

At the other extreme, a modern child growing up in the USA and surrounded by technology and our culture is taught to believe that what is exists only what technology can provide. To them video games are a type of Reality.

To be accepted as a sane and normal person we end up conforming to the standard view of Reality and making fun of those who don't agree with the norm.

A good friend of mine told me that when she was a child she used to be visited by a little girl who appeared from nowhere and played with her several times a week. The visiting girl may have been some type of spirit entity. Her uncle found out about these visitations when she was about eight years old, he told her that she shouldn't be making up things like that, since people would make fun of her. The visiting little girl soon stopped coming.

Was this just a child growing up and shedding her illusions, or was she really blocking her perceptions to conform to the accepted view of Reality in her environment?

On Using the Scientific Method to Study the Paranormal

Sometimes we label those who don't agree as having mental illness. This is often a convenient way to stick our heads in the sand and not admit that there may be things outside of our confort zone which do exist. The Soviets used to use the mental illness label for people who disagreed with their political views and then would put them into mental institutions.

To admit that unusual phenomena exist would be to admit that the personal foundation we built for ourselves throughout our life may be built upon quicksand.

Therefore, most people including most scientists are too close minded today to what Reality really may encompass to have a chance at trying to understand it.

It's time to break the mold and find new tests of objective validity to research unproven areas.

On Using the Scientific Method to Study the Paranormal

e. A Modified Scientific Method

Here is a modified Scientific Method for the 21st Century:

1) Clear the mind of all preconceptions on what is possible or not
2) Develop a Hypothesis
3) Develop an Experiment to Validate or disprove the Hypothesis
4) Determine the objectivity level of the instruments to test the theory and include subjective instruments be part of the measuring devices to get measurable results if needed
5) Do the Experiment and make the measurements
6) Publish the results with an additional scale of Objectivity relating to the instruments used

This approach also leads to a new scientific approach to peer review of experiments with this guiding principle:

The results of experiments regarding unusual phenomena should be reviewed based on the acceptance of the instruments being used, and evidence should be graded on a scale of objectivity/subjectivity too.

An example would be an experiment on whether Auras exist or not. (An Aura is thought to be energy field which psychics claim can be seen in different colors around people by those who are sensitive)

The most objective measuring instrument for validating auras would be devices, which can see infrared, ultraviolet, and other energies to see if changes corresponded to the instruments.

A more subjective but still very objective instrument overall would be to record independent testimony of a room of sensitive people as to what they thought they saw in the person's Aura (like colors, etc) during the experiment.

On Using the Scientific Method to Study the Paranormal

Significant agreement of the observers under the right conditions would provide objective evidence also.

Our individual senses should not be ignored in the advancement of Science if the proper objective controls are provided.

On Using the Scientific Method to Study the Paranormal

f. Develop A New Openness to Reality:

We have explored in this book how some events may never be measurable going by traditional objective standards because we don't have the instruments.

An equally important point is the objectivity of the observer himself.

If the observer is closed to new interpretations of scientific evidence then no matter what the results are, they will never be accepted by him or the general community which thinks the same way.

A good example are ESP experiments which have produced solid positive results with good controls thousands of time at numerous institutions for over 70 years. (Institutions with evidence include the ASPR, The Rhine Institute, and others) [W6]

Why is it then that ESP is still considered unproven by most, and people continue to waste their time trying to prove it exists instead of developing ways to use it productively?

What other answer can there be but that the majority of people are blind to accepting results which would cause them to re-interpret their structure of Reality, or at least show that there may large gaps in Science's understanding of the world.

How can this problem be resolved?

The problem is that people do not make major changes in their Reality views quickly, and don't do it without overwhelming evidence. (My own views of Reality only really started to change after a lot of major experiences, which I couldn't put in the standard framework.)
This becomes a Catch-22, because how will experimenters ever provide evidence to satisfy the majority when the evidence is

On Using the Scientific Method to Study the Paranormal

very difficult to collect and the instruments themselves are limited.

I think the answer is a new type of faith. People have to develop a new type of faith in a more open Reality, and become more open to evidence which may be unsettling but which validates the larger Reality we really live in.

This type of faith is not a religious faith, but one that seeks to build a more comprehensive understanding of Reality and our place in it than the current narrow scientific approaches.

A new class of scientist and engineer needs to be trained as well, who will look for additional tools to measure objectively; which may be built on subjective measurements by people--who if used properly can be the best measuring instruments of all.

Try taking a step towards this new openness by asking yourself a question: Are the billions of people in the world who believe in some type of God all wrong? Are we so arrogant as to say that thousands of years of spiritual
teachings are so much garbage and weren't based on any subjective evidence, which may have validity in Reality?

On Using the Scientific Method to Study the Paranormal

g. Why Many Science May Be Missing Some Of The Truth

Physics was one of the greatest areas of fundamental advancement in human knowledge in the last century. One of the greatest efforts in Physics today is to try to build a Unified Theory of all the fundamental forces of the Universe.

Theoretical Physicists today know of four types of fundamental forces:

* The Weak Force at the atomic level
* The Strong Force also at the atomic level
* Gravity
* Electromagnetism

Theories have been validated which link the Weak and Strong forces, and separately Gravity and Electromagnetism.

However, the Holy Grail of physics--a Unified Theory of everything is elusive.

The present searches by reputable scientists all assume that all of the fundamental forces are understood.

However, per my previous arguments in this book, what if there is an energy component to Reality which is not normally observed and taken into account in experiments?

If this is the case then all of the forces of the Universe are not really being addressed in the present Unified Theory efforts, and therefore all the present efforts are doomed to failure.

You can tell from the subjective experiences I've described that I think this effort will fail.

A more successful approach to a solution would be to try to account for all things which can be measured, and then try to

On Using the Scientific Method to Study the Paranormal

build a Unified Theory which is inclusive of all the forces we can objectively observe.

Dr. Stephen W. Hawking [B9] is recognized as one of the foremost Theoretical Physicists in the world today, and he holds the Isaac Newton Chair at Cambridge University in England.
In his book "A Brief History of Time", he summarizes three scenarios which must exist about the search for a Unified Theory of Physics. I quote his statement as follows:

"There seem to be three possibilities:

1) There really is a complete unified theory, which we will someday discover if we are smart enough

2) There is no ultimate theory of the universe, just an infinite sequence of theories that describe the universe more and more accurately

3) There is no theory of the universe, events cannot be predicted beyond a certain extent but occur in a random and arbitrary manner"
I would propose an addition to his statement number one as follows:
"Being smart enough means opening our minds to developing new techniques which use new instruments for observations, to better measure the full breadth of Reality."
You can also imply from my statements on Physics that other sciences may be ignoring fundamental forces and effects which will block their advancement as well.

On Using the Scientific Method to Study the Paranormal

On Using the Scientific Method to Study the Paranormal

Chapter 8: Recommendations for Action

I've spent most of this book talking philosophy and generalities.

Since I'm an engineer and businessman, I always try to come up with some practical results from what I'm doing.

Here are some thoughts on where these new open-minded approaches could take us:

1) Create new standards of scientific objectivity

As I've discussed throughout this book, new standards of scientific objectivity are need to make progress in understanding many areas presently being explored and researched

Subjective observations of the best instrument available—people—need to be controlled in such a way that they can be included in objective experimental results

2) Breaking self imposed barriers—Expand our Horizons to Explore

We are living inside self-imposed barriers of our own making.

A thousand years ago many people thought the world was flat and ships would fall off it if they ventured too far into the uncharted Oceans. When most people learned the earth was a sphere, this opened their minds to new ideas in exploration.

A few centuries ago, nobody even thought about the idea of traveling in space. Space and the Stars were part of heaven. Living people didn't travel to heaven! When Copernicus changed our view of Reality, and Newton's laws of gravity let us chart the Moon's orbit, and science and technology advanced enough it opened ---whole new places to travel.

On Using the Scientific Method to Study the Paranormal

Another analogy is the invention of the microscope. It broke the barrier to open a whole new world for observation we didn't know existed, and led to many advances.

This same "breaking of barriers" needs to be accomplished by a large number of people in the next millennium in their willingness to use innovative approaches and new types of subjectively based instruments to be able to understand many of the mysteries we have today in the context of the larger Reality.

This expansion of our vision will provide us uncounted new ideas and methods to achieve our goals with capabilities that might seem supernatural or like Magic today.

To paraphrase what a well know Science Fiction writer once said: "A sufficiently advanced technology will appear as Magic to the uninitiated."

The same principle applies to what our heirs a thousand years from now may be able to do as a normal matter of everyday activities. They may manipulate space and time and matter in ways we can't conceive.

3) Integration of Science and Spirituality

Let us start the Third Millennium by building on the base of civilization, which the last thousand years has given us.

In the Second Millennium, which is ending, humanity bootstrapped itself from a world, which was totally governed by the Spiritual view with not much in the way of science; and reached a more balanced understanding of our world through the application of scientific principles.

We should not throw out or condemn the scientific discoveries, which have led to a world engineered to standards where the majority of people have a standard of living which even kings didn't have in the past.

On Using the Scientific Method to Study the Paranormal

Keep the values and spiritual knowledge, which have been gathered over millennia and combine them with our new scientific knowledge.

The addition of two sources of knowledge isn't just like adding 1+1 to equal 2, but may be more like 1+1= much more than three.

This integration can be done by using the new observation techniques and experimental approaches discussed here to try to scientifically validate some concepts which were previously considered beyond the ability to make measurements.

It only remains for me to re-iterate that we will be much stronger and more able to explore our future together with a fusion of the scientific and spiritual approaches to discovery and collection of knowledge than with the two views conflicting with each other.

New Goals which could be achieved by these new approaches include:
* Full Unified Physics Theory taking consciousness and other energies into account

* A new way of looking at human experience as part of deciding what observations constitute mental illness, and what is just something observed outside the bounds of our present scientific understanding

* New avenues for space exploration by considering serious study of ideas proposed in science fiction as worth serious evaluation

* The start of a new breath of creativity in our civilization as people break open the limits of what is possible now and in the future

* New avenues of evolution opening for mankind as more and more people are willing to open their minds and push back the frontiers of what they used to consider fantasy or science fiction into possible realities.

On Using the Scientific Method to Study the Paranormal

On Using the Scientific Method to Study the Paranormal

Bibliography

1) Stranger than Science. Author: Edwards, Frank, 1908-1967.Edition Information: [1st ed.] Published/Created: New York, L. Stuart [1959]

2) Yoga : Discipline of Freedom : The Yoga Sutra Attributed to Patanjali
   by Barbara Stoler Miller (Translator), Patanjali, Barbara Stoller Miller (Translator), Njali Pata

3) The King James Bible-Revised Standard Version—Psalms 1:10
"Nothing under the sun is new, neither is any man able to say: Behold this is new: for it hath already gone before in the ages that were before us.

4) Psychic Discoveries Behind the Iron Curtain
Author: Sheila Ostrander and Lynn Schroeder , Published by Prentice Hall, Inc.1970

5) Books by Abraham Maslow:
Author: Maslow, Abraham H. (Abraham Harold) Main Title: Toward a psychology of being / Abraham H. Maslow. Edition Information: 3rd ed. Published/Created: cNew York : J. Wiley & Sons, c1999.
Author: Maslow, Abraham H. (Abraham Harold) Main Title: Motivation and personality. Edition Information: [1st ed.] Published/Created: New York, Harper [1954]

6) Books on St Thomas Aquinas Writings:
Main Title: Saint Thomas Aquinas, meditations for every day, adapted from the Latin of Rev. P. D. Mezard by Father E. C. McEniry. Published/Created: Somerset, O., The Rosary press, 1938.
Author: Taylor, Alfred Edward, 1869-1945. [from old catalog] Main Title: Saint Thomas Aquinas as a philosopher, Published/Created: Oxford, B. Blackwell, 1924.

7) Books on Rene Descarte's Philosophy:
Main Title: The meditations and selections from the Principles of René Descartes (1596-1650) Translated by John Veitch. With a pref., copies of original title pages, a bibliography, and

On Using the Scientific Method to Study the Paranormal

an essay on Descartes' philosophy, by L. Lévy-Bruhl.
Published/Created: Chicago, Open court Pub. Co., 1913 [cover 1912]
Main Title: The method, meditations and philosophy of Descartes; translated from the original texts, with a new introductory essay, historical and critical by John Vietch [!] ... and a special introduction by Frank Sewall .
Published/Created: Washington [D.C.] London, M. W. Dunne [1901]

8) Books on Sir Francis Bacon:
Author:Green, Adwin Wigfall, 1900- Main Title:Sir Francis Bacon, his life and works. Published/Created: Denver, A. Swallow [1952]
Author: Steegmuller, Francis, 1906- Main Title: Sir Francis Bacon: the first modern mind, by Byron Steel [pseud.] Published/Created: Garden City, N.Y., Doubleday, Doran [c1930]

9) Steven W. Hawking's Book "A Brief History of Time-From the Big Bang to Black Holes"
Published by Bantam Books April 1988

10) Author: Leadbeater, Charles Webster, 1847- [from old catalog] Main Title:The chakras, a monograph,
  Published/Created: Chicago, The Theosophical press [c1927]

On Using the Scientific Method to Study the Paranormal

Web Reference Links

Link to Site on Dr. Steven W. Hawking
  http://www.hawking.org.uk/home/hindex.html

Link to Site with Rene Descartes Info
  http://philos.wright.edu/DesCartes/Meditations.html

Link to Sites with Sir Francis Bacon Info
  http://history.hanover.edu/texts/bacon/bactable.html
  http://history.hanover.edu/texts/bacon/bactable.html

4) Link to Site on St. Thomas Aquinas
  http://www.value.net/~bromike/aquinas/thomas.html

Link To American Society for Psychical Research web site
  http://www.aspr.com/index.htm

Link to the Rhine Research Center
  http://www.rhine.org/index.html

A Link on Abraham Maslow's Needs Hierarchy
  http://www.ship.edu/~cgboeree/maslow.html

Links on Kirlian Photography
  http://www.synergy-co.com/kirlian.html
  http://www.thiaoouba.com/kir.htm

Links on Chakras
  http://www.geomancy.org/labyrint/lab-8.html

Lectures by Samuel Lentine at the US Psychotronics Associates
1982-1988   Phone (414)742-4790
  http://www.psychotronics.org

On Using the Scientific Method to Study the Paranormal

On Using the Scientific Method to Study the Paranormal

Glossary of Words and Terms Used

Aura
1 a : a subtle sensory stimulus (as an aroma) b : a distinctive atmosphere surrounding a given source <the place had an aura of mystery>
2 : a luminous radiation : NIMBUS
3 : a subjective sensation (as of lights) experienced before an attack of some disorders (as epilepsy or a migraine)
4 : an energy field that is held to emanate from a living being

ESP- extrasensory perception
perception (as in telepathy, clairvoyance, and precognition) that involves awareness of information about events external to the self not gained through the senses and
not deducible from previous experience -- called also ESP

Illusion
1 a obsolete : the action of deceiving b (1) : the state or fact of being intellectually deceived or misled : MISAPPREHENSION (2) : an instance of such deception
2 a (1) : a misleading image presented to the vision (2) : something that deceives or misleads intellectually b (1) : perception of something objectively existing in such a way as to cause misinterpretation of its actual nature (2) : HALLUCINATION 1 (3) : a pattern capable of reversible perspective
3 : a fine plain transparent bobbinet or tulle usually made of silk and used for veils, trimmings, and dresses

Meditation
1 : a discourse intended to express its author's reflections or to guide others in contemplation
2 : the act or process of meditating

Mental-(Mental Illness)
Etymology: Middle English, from Middle French, from Late Latin mentalis, from Latin ment-, mens mind -- more at MIND

On Using the Scientific Method to Study the Paranormal

Date: 15th century
1 a : of or relating to the mind; specifically : of or relating to the total emotional and intellectual response of an individual to external reality <mental health> b : of or
relating to intellectual as contrasted with emotional activity c : of, relating to, or being intellectual as contrasted with overt physical activity d : occurring or experienced
in the mind : INNER <mental anguish> e : relating to the mind, its activity, or its products as an object of study : IDEOLOGICAL f : relating to spirit or idea as opposed to
matter
2 a (1) : of, relating to, or affected by a psychiatric disorder <a mental patient> <mental illness> (2) : mentally disordered : MAD, CRAZY b : intended for the care or
treatment of persons affected by psychiatric disorders <mental hospitals>
3 : of or relating to telepathic or mind-reading powers

Metaphysics
1 b : a particular system of metaphysics
2 : the system of principles underlying a particular study or subject : PHILOSOPHY 3b
- metaphysic adjective

Objective
1 a : relating to or existing as an object of thought without consideration of independent existence -- used chiefly in medieval philosophy b : of, relating to, or being an
object, phenomenon, or condition in the realm of sensible experience independent of individual thought and perceptible by all observers : having reality independent of the
mind <objective reality> <our reveries... are significantly and repeatedly shaped by our transactions with the objective world -- Marvin Reznikoff> -- compare

Psychotronics
Psychotronics is an interdisciplinary science concerned with the interactions of consciousness, energy fields and matter. Psycho: Mind – consciousness   Tronics: Theory - physics and instrumentation

On Using the Scientific Method to Study the Paranormal

Subjective
3a c of a symptom of disease : perceptible to persons other than the affected individual -- compare SUBJECTIVE 4c d : involving or deriving from sense
perception or experience with actual objects, conditions, or phenomena <objective awareness> <objective data>
2 : relating to, characteristic of, or constituting the case of words that follow prepositions or transitive verbs
3 a : expressing or dealing with facts or conditions as perceived without distortion by personal feelings, prejudices, or interpretations <objective art> <an objective
history of the war> <an objective judgment> b of a test : limited to choices of fixed alternatives and reducing subjective factors to a minimum

Occam's razor
: a scientific and philosophic rule that entities should not be multiplied unnecessarily which is interpreted as requiring that the simplest of competing theories be preferred
to the more complex or that explanations of unknown phenomena be sought first in terms of known quantities

Philosophy
1 a (1) : all learning exclusive of technical precepts and practical arts (2) : the sciences and liberal arts exclusive of medicine, law, and theology <a doctor of
philosophy> (3) : the 4-year college course of a major seminary b (1) archaic : PHYSICAL SCIENCE (2) : ETHICS c : a discipline comprising as its core logic,
aesthetics, ethics, metaphysics, and epistemology
2 a : pursuit of wisdom b : a search for a general understanding of values and reality by chiefly speculative rather than observational means c : an analysis of the
grounds of and concepts expressing fundamental beliefs
3 a : a system of philosophical concepts b : a theory underlying or regarding a sphere of activity or thought <the philosophy of war> <philosophy of science>

On Using the Scientific Method to Study the Paranormal

4 a : the most general beliefs, concepts, and attitudes of an individual or group b : calmness of temper and judgment befitting a philosopher

Precognition
Etymology: Late Latin praecognition-, praecognitio, from Latin praecognoscere to know beforehand, from prae- + cognoscere to know -- more at COGNITION : clairvoyance relating to an event or state not yet experienced
- pre·cog·ni·tive /(")prE-'käg-n&-tiv/ adjective

Pseudoscience
: a system of theories, assumptions, and methods erroneously regarded as scientific

Psychic
1 : of or relating to the psyche : PSYCHOGENIC
2 : lying outside the sphere of physical science or knowledge : immaterial, moral, or spiritual in origin or force
3 : sensitive to nonphysical or supernatural forces and influences : marked by extraordinary or mysterious sensitivity, perception, or understanding

Psychokinesis
changing the state of a physical object, such as moving it, using only the power of the mind

Reality
1 : the quality or state of being real
2 a (1) : a real event, entity, or state of affairs <his dream became a reality> (2) : the totality of real things and events <trying to escape from reality> b : something that is neither derivative nor dependent but exists necessarily- in reality : in actual fact

Scientific Method
: principles and procedures for the systematic pursuit of knowledge involving the recognition and formulation of a problem, the collection of data through observation and experiment, and the formulation and testing of hypotheses

On Using the Scientific Method to Study the Paranormal

Subjective
1 : of, relating to, or constituting a subject: as a obsolete : of, relating to, or characteristic of one that is a subject especially in lack of freedom of action or in
submissiveness b : being or relating to a grammatical subject; especially : NOMINATIVE
2 : of or relating to the essential being of that which has substance, qualities, attributes, or relations
3 a : characteristic of or belonging to reality as perceived rather than as independent of mind : PHENOMENAL -- compare OBJECTIVE 1b b : relating to or being
experience or knowledge as conditioned by personal mental characteristics or states
4 a (1) : peculiar to a particular individual : PERSONAL <subjective judgments> (2) : modified or affected by personal views, experience, or background <a subjective
account of the incident> b : arising from conditions within the brain or sense organs and not directly caused by external stimuli <subjective sensations> c : arising out of
or identified by means of one's perception of one's own states and processes <a subjective symptom of disease> -- compare OBJECTIVE 1c
5 : lacking in reality or substance : ILLUSORY

Telepathy
The sympathetic affection of one mind by the thoughts, feelings, or emotions of another at a distance, without
communication through the ordinary channels of sensation. –

Teleportation
This is the ability to move a physical object from one location to another without traveling through the intervening space time. This is purported to be a rare mental ability which is often demonstrated in science fiction shows through the application of some futuristic technology

Will

On Using the Scientific Method to Study the Paranormal

The force of personality which is one of our core components expressed by-- mental powers manifested as wishing, choosing, desiring, or intending

On Using the Scientific Method to Study the Paranormal

Index

A Vision, 29
Actualization need, 56
American Society for
  Psychical Research, 59
Aura, 98
Buddhist, 69
Charles Darwin, 49
Copernicus, 47
Dr. Stephen W. Hawking, 88
ESP, 23, 98
Force of Will, 71
Foretelling the Future, 64
hypothesis, 43
Illusion, 98
instrumentation, 51
Kirlian Photography, 25
Kreskin, 23
Maslow's Hierarchy, 56
meditation, 27
Meditation, 36, 98
Mental, 98
Mental Illness, 53
Mental Teleportation, 65
Metaphysics, 31, 99
Objective, 99
Occam's razor, 100
Occum's Razor, 54
orgone, 69
Out of Body Experience, 64
Out of Body Experiences, 24
Philosophy, 100
Precognition, 101

Probability, 73
Pseudoscience, 48, 101
Psychic, 101
Psychic healing, 27
Psychic Healing, 64
psychically attacked, 36
Psychics, 53
Psychokinesis, 101
Psychometry, 27
Psychotronics, 99
Ptolmy, 47
Real Reality, 66
Reality, 58, 101
Rene Descartes, 44
RPI, 32
Sam Lentine, 27
Samuel Lentine, 69
Sanskrit, 31
Scale of Believability, 63, 71
Scientific Method, 48, 83
SCIENTIFIC METHOD, 43
Sir Francis Bacon, 45
St. Thomas Aquinas, 43
Subjective, 100, 102
subjective knowledge, 50
Telepathy, 64, 70, 102
Teleportation, 102
The Yoga Sutras of
  Patanjali, 31
US Psychotronics
  Association, 69
Vital Force, 27
Will, 102

www.ingramcontent.com/pod-product-compliance
Lightning Source LLC
Chambersburg PA
CBHW031441210526
45464CB00005B/2286